I0481871

Project Design Reviews

A Mentor for Successful Design Reviews

by

Denny C. Davis, PhD

Solution

Proposed Solution

Enhance, develop solution   *Evaluate, document solution*

Solution Realization

Exploration

Generate ideas   *Synthesize, select best concept*

Selected Concept

Judgment

Concept Generation

Explore needs   *Establish requirements*

Defined Problem

Problem Scoping

Challenge

Team Name: _____

Team Contact (name): _____ Phone: _____

Project Name: _____

Start Date: _____ End Date: _____

Project Design Reviews

By Denny C. Davis, PhD

Acknowledgements

The author would like to acknowledge the contributions of many others who have encouraged and enabled the preparation of this book. The importance and rigor of design reviews have been impressed upon me by Mr. Donald Dewey, formerly of The Boeing Company, and Dr. Howard Davis, faculty colleague in design education at Washington State University. The need for design assessment has been articulated well by ABET, the accrediting agency for engineering degree programs. Prominent design education scholars have emphasized the need for well-defined outcomes and feedback on performance to reach high levels of performance. I have been reminded of the need to document design thinking through the FIRST® Technical Challenge program and my high school FTC team, the Robo Raiders. I received valuable book reviews and suggestions from Dr. Susannah Howe, Dr. Will Holmes, and other capstone design colleagues. My dear wife, Irma Davis, has been my unceasing encourager as I have pursued ways to improve engineering design education. Jesus Christ, the ultimate designer, has guided me along rich paths of learning, exploring, and growing professionally in preparation for writing and publishing this book. May this book serve as a tool for enhancing design learning and increasing the rigor in design work done by engineers in the future.

©2018 by Denny C. Davis

Verity Design Learning LLC
Mascoutah, Illinois 62258
http://veritydesignlearning.com
618-566-3050

Preface

The world faces many complex technical challenges. Oftentimes, a team is formed and charged to develop an acceptable solution to a given problem in a fixed period of time. The solution must satisfy many people who have differing uses and desires for the solution. As expected, not all projects will be successful. In fact, some projects will result in loss of time and money, and some can even endanger society. Competent design professionals must be able to develop a viable solution while also ensuring public well-being and satisfaction.

A vital part of engineering education is preparing future engineers to conduct responsible engineering design and be able to defend both design processes and design products. This book is a guide and documentation template for engineering students, enabling teams to achieve and document high quality design that can pass rigorous design reviews faced in professional practice.

Chapter 1 is intended to help the reader grasp the purpose, elements, and challenges of design. This chapter provides a context for taking design seriously and building rigor into a team's approach to a design project.

Chapter 2 focuses on documentation needed for rigorous design. This chapter and the appendix provide teams basic principles and tools (including templates) for documenting both the activities and the work products of design.

Chapter 3 guides planning, execution, and documentation of problem definition for a design project. This chapter identifies what needs to be done and how it needs to be done. It prepares the team for their first design review.

Chapter 4 is a guide for concept generation and concept selection done in a design project. This chapter outlines steps to be taken, provides tools to use, and prepares the team for a review of this phase of design work.

Chapter 5 guides detailed development of a concept into a completed design solution. It guides development activity, provides tools for each step, and helps teams prepare for a rigorous final design review.

In total, this book guides teams through the design process, prepares them for design reviews, and documents the design process and products as needed to prove design competence. The completed book provides evidence of design accomplishment needed for grading and for supporting program accreditation.

Table of Contents

1. DESIGN AND DESIGN REVIEWS

What is Design?

Design is the hallmark of the engineer, the signature activity of engineers! Design is individuals working together to create new or improved products (especially technological devices, processes, or systems) that meet needs in society. Design is goal-oriented creativity under constraints of time, resources, and values important to society.

Engineers of various disciplines and roles contribute to the technologies we see and use every day. Engineers design smart devices that have the apps, performance, and look and feel we want. Engineers design, build, and test roads, bridges, buildings, vehicles, robots, recreational equipment, entertainment systems, personal care devices, power and communication systems, manufacturing systems, defense and security systems, and transportation systems. Engineers develop processes, equipment, and quality standards for food production, preservation, and preparation. Engineers create drug delivery and surgical devices, prosthetics, and patient care equipment. Engineers locate, extract, and refine natural resources for use in society. Applying knowledge of natural sciences, modern technologies, and social sciences, engineers convert materials and energy into new and improved products for the benefit of society.

Design activities addressing a given need can produce a range of solutions, each one having a different match to the need and potential for delivering benefits to society. Although no single solution is the "right answer", some are definitely better than others. The challenge in design is to produce solutions that fit the needs well, provide real value, are durable, and are beneficial to society. This is creativity under constraints.

THINK: Context for Design
A design project should yield a solution that meets needs of the intended audience. Who might be the audiences for a drinking water filtration system for rural Malawi, Africa? What dangers or undesirable costs might result if this design is done poorly?
Audiences:
Dangers:
Costs:

Design Process

Engineers use a design process to create a solution that meets needs of individuals and/or groups. Design is activity focused on producing the desired product, moving from abstract to very specific understanding of the solution. Phases or stages of design include:

    (a) **Problem Scoping** - vague challenge → well-defined problem
    (b) **Concept Generation** - defined problem → solution concept
    (c) **Solution Realization** - concept → proven design solution.

Figure 1 shows a design process, beginning at the bottom with a problem or challenge and progressing upward to the design solution. The design process uses both exploration (creativity) and judgment (decision making) in developing a solution. Problem scoping activities (defining the problem) look broadly to explore the problem and its context, then they judge (select) and synthesize needs to define requirements (specifications) the solution must meet. Selecting a "best" concept uses exploration in identifying ideas, then it judges these to select the one best fitting the solution requirements. A proposed solution results from exploring and refining the concept to a solution, then judging the extent to which it meets requirements. The design process is frequently interrupted by returns to earlier steps to improve the defined problem, concept, or emerging solution. The process shows that designers must be creative, able to evaluate, and able to solve problems that likely change over time.

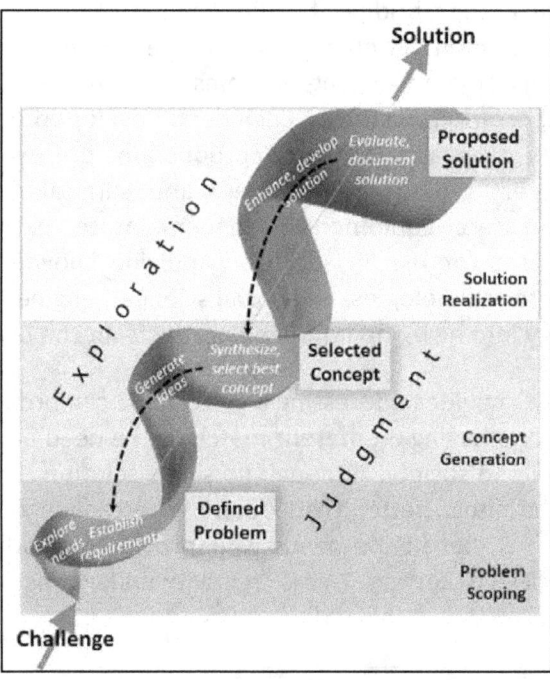

Figure 1. Design process exploring and judging to yield a solution

THINK: Design Process
A design solution must meet established requirements. What will happen if the requirements are incorrect? What will happen if a solution is not evaluated thoroughly?
Incorrect requirements:
Poor evaluation:

Design Reviews

The quality of a design solution is determined by the effectiveness of the many steps leading to the solution. Errors made in any design step will affect the work in subsequent steps. Any errors occurring early but not detected until much later can require massive amounts of rework, delaying the project and increasing its costs. Any design errors that remain undetected during design can cause significant costs in user dissatisfaction, product failures, safety violations, injuries, and even loss of life. Therefore, engineers must ensure high quality in their design processes.

Design reviews are formal examinations of design processes and products to ensure their adequacy at critical points in the design process. The Food and Drug Administration (FDA) defines a design review as "a documented, comprehensive, systematic examination of a design to evaluate the adequacy of the design requirements, to evaluate the capability of the design to meet these requirements, and to identify problems."[1] Engineers working on designs in food, pharmaceuticals, medical, aerospace, or other sectors where human life is at stake must comply with strict design controls that require periodic rigorous design reviews.

Based on the FDA definition of a design review, the following questions emerge as a framework for design reviews:

1. How adequate are the design requirements? How well do they represent needs of users? How well do they support design decisions and testing? How well are they reviewed and updated as conditions change?
2. How adequately does the design meet the established requirements? How effectively are requirements used in making design decisions? How well has the design been evaluated for achievement of requirements? How well has the design been checked by users to ensure that it meets their needs?
3. What problems have been identified? What steps are required to address problems identified? What must be done before the design may advance?

THINK: Rationale for Design Reviews
A design review is used to check adequacy of design at critical points in the design process. Why should design reviews be conducted multiple times, not only at the end of a project? Why should design documentation be reviewed as part of a design review?
Multiple reviews:
Documentation:

---

[1] US Food and Drug Administration. *Medical Device Quality Systems Manual: A Small Entity Compliance Guide.* 2012. Silver Spring, MD, section 320.3(h)

Design Review Timing

Design reviews should be conducted at important milestones in the design process to ensure that achievements justify and provide conditions that support the next phase of the design project. As shown in Figure 2, design reviews logically occur at the end of the **problem scoping**, **concept generation**, and **solution realization** periods. At these points, a review can examine both the end product of that phase and the process used to create the product. As indicated by curved arrows, a design review may require a repeat of earlier steps, which may even cascade back to much earlier steps.

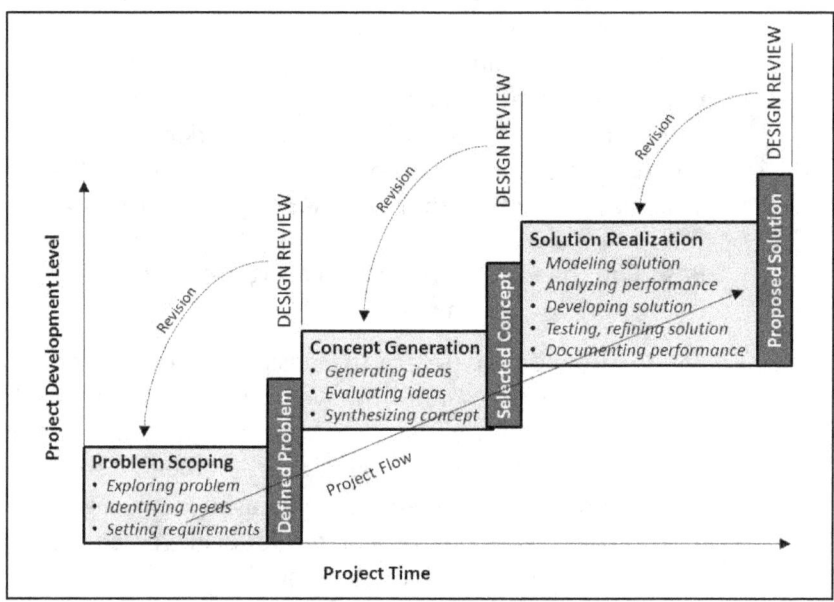

Figure 2. Design reviews along the design process

Design reviews at these milestones can yield multiple benefits:

- Project team approved to advance to the next design phase,
- All important elements of design process and products reviewed,
- Design team receives feedback to guide improvements to design,
- Project progress understood by designers and audience, and
- Documentation of process and product checked for adequacy.

THINK: Early Use of Design Reviews

Early design reviews are conducted upon completion of a significant early design product. Why is the product of an early design phase important? Why is the (already completed) early process important to review?

Product review:

Process review:

4

Dual Foci of Design Reviews

Design reviews prescribed in design controls regulations call for checking two types of achievements, as illustrated in Figure 3:

- **Verification**: Assurance that the design team has satisfied all design requirements (or specifications) that were <u>anticipated</u> and defined for an acceptable solution, and
- **Validation**: Assurance that the design product meets <u>actual</u> user needs in the varied applications in which it is employed by users.

The design review looks at how well the design activity is achieving expectations for a stage of design. **Verification** checks the effectiveness of the team's design activities to accomplish what they targeted, i.e., meeting solution specifications. For example, the team may have produced a device that is 99.5% effective in separating two types of seeds in the laboratory. On the other hand, **validation** checks to see that the team delivered what actually works in the conditions under which it is used. Actual use conditions may not be the same as conditions defined in the solution requirements. For example, requirements may have ignored conditions of static electricity under which the separator must perform.

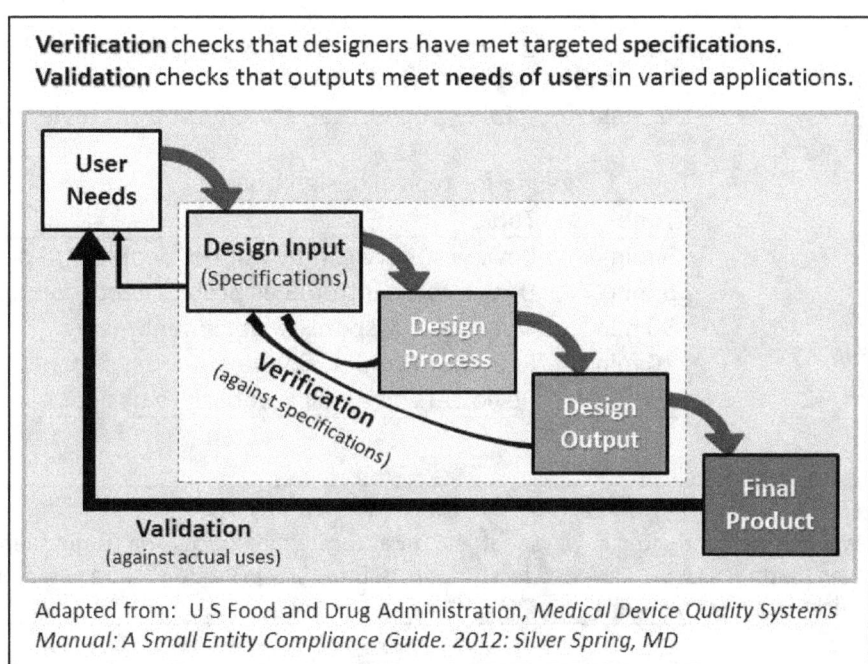

Verification checks that designers have met targeted **specifications**.
Validation checks that outputs meet **needs of users** in varied applications.

Adapted from: U S Food and Drug Administration, *Medical Device Quality Systems Manual: A Small Entity Compliance Guide. 2012: Silver Spring, MD*

Figure 3. Verification and validation used in design reviews

THINK: Verification and Validation
Design reviews require both verification and validation. Why are both required to ensure design adequacy?

Design Review Protocols

For a design review to provide greatest value, review protocols must be understood and followed by all participants. Each person must prepare for their respective part and maintain focus on the goals of the review.

For a design project in an academic setting, design review should be conducted by the supervising instructor and individuals with expertise or project investment that justifies their participation in questioning. All students with major project responsibilities or vital knowledge should be present for review questioning. Individuals most familiar with project documentation must be present to support design team claims.

Design reviews should be conducted during a designated time period (e.g., 1 hour for a final design review), with an agenda known in advance. Review questions should be given to all parties in advance to allow everyone adequate time for preparation. As appropriate, students might be asked to submit written responses to questions prior to the review. This enables reviewers to give adequate time to issues of concern so they will be able to give more helpful feedback for improving the design.

A typical design review agenda might be as defined in Table 1. This agenda provides the design team immediate oral feedback on the review and sets the stage for a follow-up written statement from reviewers.

Table 1. Agenda for typical design review

Time	Topic
5 min	Review supervisor introduces people and process
5 min	Design team introduces project need, solution, benefits
30 min	Questions, responses, document review
10 min	Reviewers confer, define outcome and initial feedback
10 min	Reviewers give oral feedback to design team

THINK: Design Review Preparation
Design reviews must be conducted in ways that lead to accurate judgments and useful feedback. What are the openness issues that need to be discussed by students before they prepare for a review? What preparation will take the most design team effort?
Openness issues:
Preparation effort:

Discuss the following questions among team members to probe the understanding of individual members and your team as a whole. Write your responses below, noting differing opinions where appropriate.

1. IDENTIFY THE BENEFITS YOUR TEAM WOULD HOPE TO GAIN FROM A DESIGN REVIEW.

2. WHAT FEARS DOES YOUR TEAM HAVE ABOUT PARTICIPATING IN A DESIGN REVIEW?

3. HOW MIGHT YOU ADDRESS SOME OF THESE FEARS, TO TURN THEM INTO OPPORTUNITIES?

4. WHAT ASPECTS OF YOUR DESIGN **PRODUCTS** WILL BE MOST DIFFICULT TO DOCUMENT? WHAT CAN YOU DO TO ADDRESS THIS?

5. WHAT ASPECTS OF YOUR DESIGN **PROCESS** WILL BE MOST DIFFICULT TO DOCUMENT? WHAT CAN YOU DO TO ADDRESS THIS?

6. WHAT CAN YOU DO TO MAKE YOUR NEXT DESIGN REVIEW OF GREATEST VALUE TO YOU?

7. WHAT CAN YOU DO TO MAKE YOUR NEXT DESIGN REVIEW OF GREATEST VALUE TO YOUR REVIEWERS?

8. WHAT DO YOU NOT UNDERSTAND ABOUT DESIGN REVIEWS? HOW CAN YOU GET ANSWERS YOU NEED?

2. DOCUMENTING DESIGN REVIEWS

Documentation Essentials

Documentation of design activities and outcomes is important to every design review. As noted earlier, documentation is a key part of the FDA definition of design reviews: "a **documented**, comprehensive, systematic examination of a design . . . ." Design review documentation includes two parts:

1. Documentation of earlier design activities and products referenced in the team's responses to design review questions, and
2. Documentation of outcomes of the design review that record the reviewers' assessment of design adequacy and required next steps.

This section describes essentials of good documentation, especially as related to design reviews.

Documentation Principles

Technical writing is done for the purpose of communicating factual information to be understood without uncertainty, free of errors that might cast shadows on its credibility. Criteria used for high quality technical writing include:

- Substance – Content addresses important issues adequately to achieve the purpose of the writing
- Organization – Structure of sections and paragraphs present logical flow and development of thoughts
- Mechanics – Sentence structure, word choice, spelling, grammar, and punctuation give clarity and credibility
- Evidence – Accuracy of statements, supporting arguments, and data support claims

Engineers and technical writers need to appreciate the importance of high quality technical writing because their writing opens or closes doors to employment, advancement, and having desired impact on readers.

THINK: Technical Writing
High quality technical writing communicates effectively. What impact might the following have on a person reviewing your design report?
Many errors in mechanics:
Use of slang:
Confusing sentences:
Lack of supporting evidence:

Documenting Processes

Design teams use familiar as well as uncommon processes in their design work. These processes must be documented to enable others to check them for appropriateness or to duplicate them in the future. If processes cannot be repeated based on documentation, the documentation is inadequate. Therefore, process documentation is important to design work and to design reviews.

Processes may be presented as flow charts, lists of steps, or narrative descriptions, perhaps in a Methods section of a report. For added clarity, multiple formats may be used, such as a flow chart accompanied by a narrative description of steps. Because a process does not exist for its own sake, but is expected to produce results, process documentation probably includes the following types of information:

- Date and time – When activities are conducted
- Participants and roles – Who is responsible for which parts
- Procedures – Specific steps followed (reference standard methods)
- Influences – Equipment, software (version), materials used; context or environmental conditions of relevance
- Results – Data collected, observations, outcomes, experiences
- Interpretation – Explanation of findings, conclusions, next steps

For example, a team may document its brainstorming process to generate ideas in the following way. They first define the date and time, location, purpose, participants (recorder, facilitator), and ground rules used for brainstorming. Next they record all ideas generated. Finally, they reflect on the adequacy of the brainstorming effort and record any next steps they plan to use to obtain more or improved ideas.

THINK: Documenting Processes
Process documentation is a required part of design documentation. Describe steps you might use to obtain a consensus decision on what CAD software your team will use in its design work.
Step 1:
Step 2:
Step 3:
Step 4:
Step 5:
Step 6:
Step 7:

Documenting Products

Design products must be documented at various stages of a design project. In fact, a design history should be created to archive design product states at different times over the duration of the project. The documentation at any one time must define the design product's state well enough so that key features are understood correctly. Product documentation may need to support any of the following actions:

- Building or prototyping components or the entire product
- Conducting analysis of the product with regard to cost, function, durability, or impacts on surroundings
- Communicating to others the product's state of development
- Evaluating compliance with requirements, specifications, or standards

This suggests the following criteria for high quality product documentation:

- Completeness – Describes all relevant parts of the product
- Specificity – Defines details with specificity (including tolerances) and clarity needed to conduct appropriate follow-up actions
- Relationships – Presents relationships among parts and with other entities to understand interfaces and the whole
- Operation – Describes how the product operates

Documentation of design products may take different forms depending upon the type of product. In any case, adequate documentation gives a representation of the product that communicates all important aspects of the product at this stage of development. Examples of product documentation formats are listed below.

- CAD drawings of individual parts and assembly procedures
- Photos of prototype in various positions of movement
- Sketches with critical dimensions and tolerances specified
- Printed circuit board layout drawings for each separate layer
- Formulation and mixing instructions for a proposed product
- Map of construction site with details on referenced sheets
- Computer code documented, dated, and credited

THINK: Documenting Design Products
Documenting design products is a required part of design documentation.. Describe the details you would need in design drawings of a spring-driven car if you were to build it correctly from this documentation.
Car body:
Power system:
Steering:
Other:

Documenting
Learning

An important part of design projects is the learning gained. Learning may include increased understanding of the problem addressed, knowledge about viable solutions, understanding of concepts, or discernment about profitable approaches to design. Documenting this learning can prove valuable in design reviews and in demonstrating your capabilities for performing high quality design.

Documentation of learning should occur throughout a design project at the conclusion of planned thought-inspiring project events, and also when "aha" moments occur unexpectedly. Project-related events that should produce significant documented learning include design reviews, meetings with project sponsors, meetings with the instructor, and the completion of major reports or presentations. Additional unplanned documentation of learning should occur when questions are asked of the instructor, teammates, or project advisors and new insights are gained.

Learning is vital to your development of expertise. Development of expertise requires that new learning be organized and connected to previous learning, and that new learning be applied in varied situations where feedback can be obtained.[2] Your documentation of new learning can best support building expertise when you include:

- Date and setting in which learning occurred
- Statement of new facts or concepts or insights gained
- Explanation of how the new relates to previous understanding
- Description of where the new can be applied and where it cannot
- Reflection on how this learning occurred and how this process relates to future learning

THINK: Documenting Learning
Documenting learning creates new knowledge assets for you and your team. What barriers commonly prevent you from documenting your learning? How might you overcome these barriers?
Barriers:
Strategy to overcome:

---

[2] Litzinger, TA et al., Engineering Education and the Development of Expertise. Journal of Engineering Education, 2011: 100(1).

Document Preparation for Design Reviews

An important part of preparing for design reviews is the organization of design documentation to support a successful review. A design team must ensure that supporting materials address expected review questions and that these materials are readily accessible during the review.

Types of Materials

Materials will be needed to respond to two types of questions:

(a) What was done or how was it done? (process questions)
(b) What was achieved, produced, or found? (product questions)

Materials addressing process-type questions or histories of product states may have been generated over an extended period of time, and therefore, may be dispersed widely in project documentation. Portions may be text documents, photo images, charts, etc. that are not easily found or combined when needed to support a review question response.

Materials answering product-type questions often include lists, drawings, citations, physical models, audio-visual recordings, or other artifacts that may be available, but the challenge remains in synthesizing different types of materials to present a coherent description.

Organization of Materials

Responses to questions should provide the larger context and the detail desired by reviewers. Thus, a team should organize materials to first show the "big picture" and then provide supporting details as needed.

Big Picture. Prepare an overview document that answers the question without details: a table, chart, list, low-detail image, or summary statement. This serves as an introduction or cover sheet to other materials.

Backup Details. Compile additional documents or other artifacts that show important details of relevance to the question. These items should be indexed or linked from a table that identifies item, location, and details addressed in each (for quick reference).

THINK: Preparing for a Design Review

A design team must enter a design review with appropriate supporting documentation at hand. What type of "big picture" item would you use to explain how your team selects the best concept from a set of proposed concepts? How would you show details of this process?

Big picture:

Details:

Team Discussion: Documentation for Design Reviews	To deepen your understanding and enhance your preparation for design reviews, discuss the following questions within your team. Write your responses below.

1. WHY IS DESIGN PROJECT DOCUMENTATION DEMANDED BY REVIEWERS IN A DESIGN REVIEW?

2. WHY MUST THE OUTCOMES OF A DESIGN REVIEW BE DOCUMENTED AND KEPT AS PART OF PROJECT RECORDS?

3. WHAT PROBLEMS MIGHT BE CAUSED BY INCOMPLETE OR INACCURATE RECORDS OF TESTING DONE ON A DESIGN?

4. IF DESIGN RECORDS CONTAIN POOR WRITING AND MANY ERRORS, HOW MIGHT THIS AFFECT AN EVALUATION OF THE DESIGN WORK?

5. IF DESIGN DRAWINGS LACK KEY INFORMATION NEEDED TO BUILD THE PRODUCT, WHAT PROBLEMS COULD ARISE WHEN THE PRODUCT IS BEING TEST MARKETED?

6. WHEN PREPARING FOR A DESIGN REVIEW, WHAT TYPE OF OVERVIEW DOCUMENT MIGHT YOU CREATE TO SHOW HOW THOROUGHLY YOU SEARCHED FOR DESIGN IDEAS?

7. WHAT MATERIALS MIGHT YOU HAVE AVAILABLE TO SUPPORT AN OVERVIEW DOCUMENT PREPARED TO ADDRESS QUESTION 6 ABOVE?

8. HOW MIGHT YOUR TEAM ORGANIZE ITSELF TO ENSURE THAT DOCUMENTATION FOR A DESIGN REVIEW WILL BE WELL PREPARED?

3. PROBLEM SCOPING DESIGN REVIEW

Components of Problem Scoping Review

The problem scoping design review is conducted after the design team prepares its definition of the problem and establishes requirements for a design solution to satisfy people who have an interest in the solution. Figure 4 shows the place of this design review on the overall project timeline. The problem scoping process and defined problem are expanded in this figure to show the components of each.

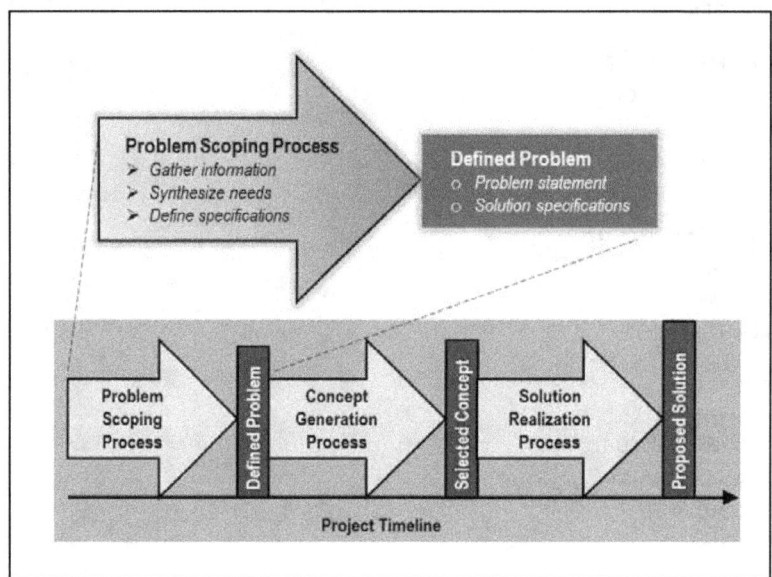

Figure 4. Components of problem scoping process and products

The problem scoping design review examines both the problem scoping process and the defined problem produced by this design phase.

THINK: Problem Scoping Review
The problem scoping design review checks the quality of the problem scoping process and of the defined problem. Why is the process checked rather than only the defined problem? What qualities of the defined problem are necessary before it is adequate?
Why process:
Qualities of defined problem:

Adequacy of Problem Scoping Process

A rigorous review of the problem scoping process ensures that problem definition and solution requirements are derived from suitable information and reasoning. Flawed understanding of the problem can occur when important information sources are neglected, untrustworthy or biased sources are given credibility, or similar problems and solutions are not considered. In addition, effective methods must be used to compile, sort, combine, and prioritize the needs identified. Moreover, important needs must be converted into solution specifications that guide design work and are used to evaluate design ideas. A final check is to validate with stakeholders that the derived solution specifications fit their real needs.

These considerations are used to define questions and scoring factors for the problem scoping process review, as presented in Table 2. Questions focus on the adequacy of the information gathering for definition of needs and on the derivation of explicit solution specifications from these needs.

Table 2. Questions and scoring factors for problem scoping process review

Question	Scoring Factors
How adequate is information gathering for understanding the design problem?	• Breadth and relevance of sources used • Credibility of sources • Breadth of needs identified • Inclusion of relevant regulations and standards
How adequate is synthesis of needs and derivation of solution specifications?	• Clarity and consistency of process used • Inclusion of synthesis, prioritization, and quantification • Validation of specs with stakeholders

As implied by Table 2, the design team should organize the following design records to support a review of their problem scoping process:

- List of information sources used and full citation for each
- List of needs identified and prioritized synthesized list of needs
- List of steps or flow chart, with examples, showing process for deriving and validating solution specifications

---

THINK: Problem Scoping Process Review

The problem scoping process investigates the problem and derives requirements or specifications for a good solution. Why might international standards be an important source for understanding needs for a handheld electronic device being designed? Why is prioritization of needs important to the problem scoping process?

International standards:

Prioritization of needs:

Adequacy of Defined Problem

A rigorous review of the defined problem ensures that the design team and stakeholders agree on the scope of the problem being addressed and the specific requirements (specs) to be met by the design solution. Erred scope or focus of the design effort can waste time and yield a design solution of limited value. On the other hand, a compelling problem statement can motivate the team and attract resources for a successful design project. If solution specs are defined poorly, the resulting design solution may solve the wrong problem. Correcting misconceptions at this early stage of design saves money, time, and reputations of all involved.

Questions and scoring factors for a review of the defined problem are given in Table 3. Questions on the problem statement examine the focus, scope, and compelling nature of the problem and its solution. Questions on solution specs give attention to clarity and testability of specs, as well as their ability to allow creativity within constraints of the problem.

Table 3. Questions and scoring factors for defined problem review

Question	Scoring Factors
Briefly describe the problem you address, your envisioned solution, and benefits you expect.	• Clarity, urgency, relevance of need • Solution achievability and fit to need • Potential for benefit to users and more broadly
Illustrate with examples the breadth and quality of specifications you derived for your design solution.	• Clarity, abstractness, and testability of solution specifications • Specs appropriately span function, financial, technical, and social issues
Describe how your defined problem will be used and might change over time.	• Use in design activities and decisions • Change in response to new knowledge

The design team should assemble the following design records to support the review of their defined problem:

- Written problem statement, about one paragraph in length
- List of solution specifications with targeted states to be achieved
- Next steps, illustrating how the defined problem will be used

THINK: Defined Problem Review
The defined problem describes the problem addressed and requirements for a good solution. How might a vague problem definition lead to excessive design costs? What makes a problem statement compelling?
Cost of vagueness:
Compelling problem:

Summary of Materials for Problem Scoping Design Review

The following is an example summary of materials prepared to support a problem scoping design review. The notebook references identify where in the design notebook to find details of the supporting materials.

Problem Scoping Process

Information Gathering

Type	Description	Reference	Notebook
Customer interviews	E.J. Jones, L. Sanchez (09/10/2014)	Audio recorded by E. Brown	pp. 41-43
Patent search	Mechanisms for grasping spheres	Patent numbers: US8231158	p. 51
Market research	Review of hand-held mixers	Consumer Reviews 10(2):100	p. 57

Needs Identified *(summarized on notebook page 77-79)*

Type	Description	Source	Importance
Safety	Avoid cuts from sharp edges	R. Stanislaw (user)	H
Durability	Last longer than Brand X	L. Sun (owner)	M
Appearance	Look rugged	K. Kurtz	M

Process for Derivation of Specifications *(see notebook pp. 83-88)*

Step	Description
1	Compile needs into similar types
2	Combine similar needs
3	Prioritize needs
4	. . .

Defined Problem

Problem Statement *(see notebook pp. 45, 97)*

```



```

Solution Specifications *(see notebook pp. 99-104)*

Description	Targeted State	Importance
Overall maximum dimensions (in)	12W x 15L x 18H	H
Time to complete one cycle (sec)	15	M
Electrical circuitry certification	UL standards	M

How to Score Problem Scoping Review

A problem scoping design review should give the design team feedback on both the design work accomplished and their documentation of this work. Reviewers use the score sheet on the following page to conduct the problem scoping design review. The sheet contains questions for the review, offers items for check-off on content, and defines scoring scales for both design content and documentation associated with each response.

Guidelines for Scoring Problem Scoping

A score of 1 to 3 is awarded for the students' oral response to each design review question. A score of 0 to 2 given for supporting documentation communicates that poor documentation is worth nothing. It also leads to a maximum of 5 possible points for each question, simplifying scoring calculations. Guidelines for conducting the review are stated below.

Questioning. Read the stated question. Ask the students to respond to the question and show documentation to support their response.

Note taking. Listen to responses to see if students include elements listed on the score sheet for that question. On the sheet, check items addressed and note relevant details included in the response.

Scoring. Using the scoring scale for each question: a score of 1 to 3 for the oral response and a score of 0 to 2 for supporting documentation.

Giving Feedback on Problem Scoping

At the end of the problem scoping design review, reviewers should provide feedback similar to that defined in Table 4. Outcome options include (a) acceptable, (b) acceptable with minor revisions, and (c) unacceptable.

Table 4. Outcomes from problem scoping design review

Decision/Scoring	Subsequent Actions Approved
Acceptable: All scores at 3 or above	Proceed to concept selection; no revision required
Acceptable with minor revision: Most scores at 3 or above	Proceed to concept selection after revising response or documentation to make all scores acceptable
Unacceptable: Many scores below 3	Revise problem scoping content and/or documentation; undergo another problem scoping review before proceeding

---

THINK: Feedback on Problem Scoping Review

Feedback on problem scoping can guide improvement and can approve continuation of design. What will happen to your team if you do not receive an "acceptable" score? How does this affect your stakeholders?

Impact on team:

Impact on stakeholders:

Scoring Sheet for Problem Scoping Design Review

INSTRUCTIONS: Ask students questions listed in column 1. For each question, check off items in column 2 if they are addressed adequately by students. For the question, assign one score for the response and one for documentation supporting the response. Tally scores and circle the outcome corresponding to these scores.

Problem Scoping Process

Probing Question	Addressed in Response	Response Score			Documentation		
		1 *Unacceptable*	*2* *Acceptable*	*3* *Outstanding*	*0* *Little*	*1* *Marginal*	*2* *Complete*
List **information sources** you used to identify varied needs and expectations for your solution.	□ Possible users: □ Other people: □ Patents, copyrights or products: □ Reports or studies: □ Rules, policies, standards:	Few, narrow, or unreliable sources	Moderate variety and credibility in sources	Widely varied, necessary, and authoritative sources	Very little record; unclear on sources	Sporadic, some well-defined sources	Complete, dated, full citations on sources
Give examples that show the **different types** of needs and expectations you identified.	□ Physical characteristics: □ Functional performance: □ Financial constraints: □ Building, servicing, disposal issues: □ Human, safety, societal concerns:	Few, narrow, unclear, or unbelievable	Moderate variety, clarity, importance, credibility	Comprehensive, clear, important, authoritative	Very little record, unclear needs	Mixed record, some clear needs	All clear, well-defined needs
What **regulations** or **standards** apply to development or use of your solution?	□ Health or safety: □ Environmental: □ Manufacturing: □ Other:	Missing important ones	Suitable but not complete	Thorough, correct, valuable	No record of attempts	Vague, weak definitions	Fully referenced & quoted
What **process** (steps) did you use to define solution requirements or specifications? Show an example.	□ Synthesis of needs: □ Prioritization of needs: □ Conversion to specifications: □ Selection of targeted state/value: □ Validation of specs with stakeholders:	Unclear or very incomplete process	Moderately complete and clear process	Very complete, clear, and rigorous process	Very little record; unclear process	Sporadic record, parts of process	Complete, record of process & issues

Defined Problem

Probing Question	Addressed in Response	Response Score			Documentation		
		1 *Unacceptable*	*2* *Acceptable*	*3* *Outstanding*	*0* *Little*	*1* *Marginal*	*2* *Complete*
In 30 seconds or less, summarize the **need** you are addressing, your envisioned **solution**, and **benefits** it will deliver.	□ Clear, compelling need: □ Need connects with audience:	Unimportant or unclear need	Moderate need, clearly stated	Great need; urgent; motivates action	Very little record; unclear	Single acceptable entry	Clear, refined, prominent
	□ Solution fits stated need: □ Solution is achievable:	Vague or unlikely solution	Relevant, maybe feasible solution	Great solution; very likely achievable	Very little record; unclear	Single acceptable entry	Clear, refined, prominent
	□ Promises real benefits to users: □ Promises potential broader benefits:	Unclear or unlikely benefits	Probable benefit to users, others?	Likely big benefits to users & others	Very little record; unclear	Single acceptable entry	Clear, refined, prominent
Show examples to illustrate the **breadth** of your solution specifications.	□ Physical or function: □ Production or service: □ Financial or value: □ Safety or societal:	Very narrow; important types missing	Important types included, some barely adequate	Comprehensive inclusion of all important types	Very little record, not coherent	Mixed, some parts coherent	All clear, complete, coherent
Show two of your specifications that are **testable** and **central** to user expectations.	□ Clear and relevant: □ Allows creativity: □ Testable: □ Conforms to needs and uses:	Marginally understandable; few testable	Understandable; most relevant and testable	Very clear; vital to success, testable; allow creativity	No record of attempts	Vague, weak definitions	Fully referenced & quoted
Explain how you expect specs will be **used/changed** in future design effort.	□ Criteria for design decisions: □ Basis for verifying solution: □ Evolve as project progresses:	Vague use in design effort	Clear use in design decisions	For decisions & evaluation; refine by new information	No context given for specs	Tied to design process	Focus for reviews; revised

Problem Scoping Design Review **Outcome**

Suggested Score	30-50	25-29	10-24
Outcome	Accept as is	Accept with revisions	Revise and re-review

To deepen your understanding and enhance your preparation for a problem scoping design review, discuss the following questions within your team. Write responses beside each question for later reference.

1. WHY IS A PROBLEM SCOPING DESIGN REVIEW IMPORTANT FOR YOUR PROJECT?

2. WHAT SPECIFIC THINGS WOULD YOU LIKE TO LEARN FROM A PROBLEM SCOPING DESIGN REVIEW?

3. WHAT STRENGTHS DO YOU FEEL YOUR TEAM HAS DEMONSTRATED DURING YOUR PROBLEM SCOPING?

4. WHAT SPECIFIC THINGS SHOULD YOUR TEAM HAVE DONE BETTER DURING PROBLEM SCOPING?

5. WHAT DESIGN RECORDS DO YOU HAVE TO SUPPORT YOUR PROBLEM SCOPING DESIGN REVIEW? HOW WILL YOU USE THEM IN THIS REVIEW?

6. WHAT PROBLEM SCOPING ACTIVITIES OR OUTPUTS ARE **NOT** DOCUMENTED WELL? WHAT CAN BE DONE TO FIX THIS PROBLEM?

7. GIVE EXAMPLES OF SOLUTION SPECIFICATIONS THAT ARE SPECIFIC ENOUGH TO BE TESTED.

8. GIVE EXAMPLES OF SPECS THAT ARE TOO RESTRICTIVE TO ALLOW NEEDED CREATIVITY. HOW MIGHT THEY BE REVISED TO ALLOW MORE CREATIVITY?

4. CONCEPT GENERATION DESIGN REVIEW

Components of Concept Generation Review

The concept generation design review is conducted after the design team selects its concept to be developed into a full design solution. Figure 5 shows the place of this design review on the overall project timeline. The concept generation process and selected concept are expanded in this figure to show the components of each.

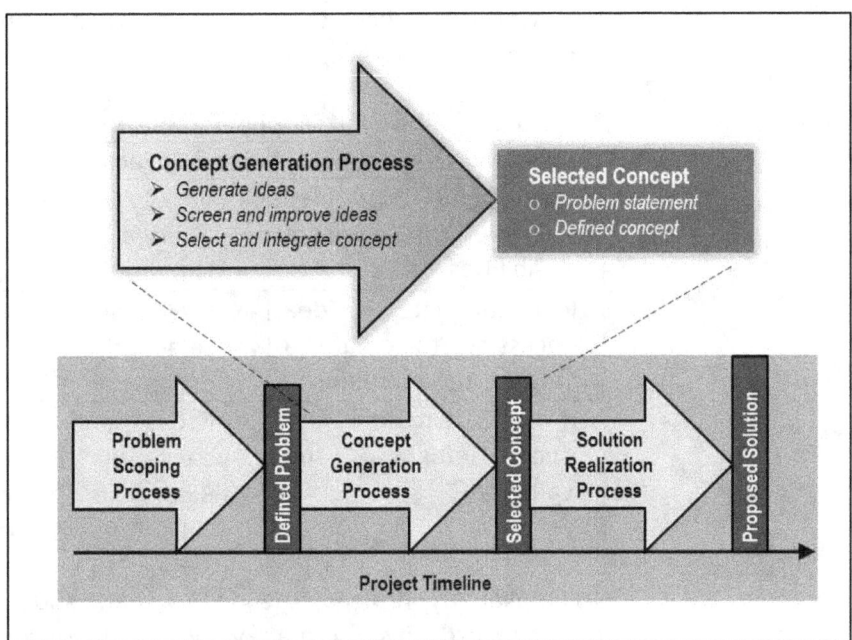

Figure 5. Components of concept generation process and products

The concept generation design review examines both the concept generation process and the selected concept created in this design phase.

THINK: Concept Generation Review
The concept generation design review checks the quality of the concept generation process and of the selected concept. Why is the process checked and not just the selected concept? What characteristics of the selected concept are important for it to be judged adequate?
Why process:
Characteristics of selected concept:

Adequacy of
Concept Generation
Process

A rigorous review of the concept generation process ensures that the selected concept is derived from suitable idea generation, idea selection, and concept synthesis activities. An inappropriate concept can be defined when important idea sources are missed, poor screening and selection methods are used, or components are ineffectively integrated into a solution concept. A final check is to validate the concept with stakeholders to ensure that it has potential to meet their needs in their applications.

These considerations are used to define questions and scoring factors for the concept generation process review, as presented in Table 5. Questions focus on the adequacy of the idea generation activities and on selection and synthesis processes used to create a solution concept that meets solution specifications.

Table 5. Questions and scoring for concept generation process review

Question	Scoring Factors
How adequate is idea generation of concept possibilities?	• Variety and relevance of methods used for idea generation • Number and value of ideas generated
How adequate is idea processing to create a match to requirements?	• Clarity and appropriate use of specs in selection, evaluation, and improvement process
How adequate are concept integration and evaluation?	• Focus and systems perspective in concept integration • Rigor in concept evaluation and validation with stakeholders

As implied by Table 5, the design team should organize the following design records to support a review of their concept generation process:

- List of idea generation methods and sources used, with details of substantive processes used
- Compilation of ideas generated from all sources
- Description of process used for evaluating ideas, with example
- Description of process used for synthesizing "best" concept
- Process used for verifying and validating solution concept

THINK: Concept Generation Process Review
The concept generation process seeks to find and develop ideas into a useful solution concept. Why are multiple methods used for generating ideas? How are solution specs used in selecting the best ideas?
Multiple methods:
Solution specs:

*Adequacy of
Selected Concept*

A rigorous review of the selected concept ensures that the design team and stakeholders agree that the concept meets the specifications and user expectations for the design solution. Note that the problem definition may be revised as the design progresses and more information becomes available. A selected concept that does not satisfy solution specs will yield a design solution of limited worth, wasting valuable time and resources.

Questions and scoring factors for a review of the selected concept are given in Table 6. Questions on the defined problem ensure suitable targets for the solution, while questions on the selected concept focus on its satisfying these requirements.

Table 6. Questions and scoring factors for selected concept review

Question	Scoring Factors
Briefly describe problem and explain how it has changed or been confirmed.	• Clarity of need and recentness of its definition
Describe the solution concept's essential features and functions and how well they meet important specs.	• Clarity and relevance of concept features and functions • Breadth and credibility of evidence that concept meets specs
Describe how your selected concept will be improved and advanced to a detailed design solution.	• Accuracy and progress seen in concept assessment • Value of plans for advancing concept to completed solution

The design team should assemble the following design records to support the review of their selected concept:

- Brief problem definition with important specs and revision dates
- Description of solution concept with key features and functions
- Summary of evidence verifying that specs are met by the concept
- Self-assessment of selected concept: strengths, concerns, and adequacy of progress
- Plans for advancing concept to a full solution: principal activities to be conducted to achieve various goals

THINK: Selected Concept Review
The selected concept represents your team's current vision of a design solution. Why is it important to test the concept against specs at this time? What value is added by the team assessing their progress to date?
Concept testing:
Assessing progress:

Summary of Materials for Concept Generation Design Review

The following two pages contain an example summary of types of materials used to support a concept generation design review. Notebook pages identify locations of detailed support materials.

Concept Generation Process

Idea Generation Methods and Sources

Source/Type	Method Description	Results	Notebook
V. Strauss (materials processing expert)	Interview: semi-structured questions, audio recorded	Methods for joining delicate parts	p. 91
Bolts Today (catalog)	Search for fastener types	Eco-friendly fastener	p. 129
Team brainstorming	Initial idea generation, ideal solution, worst solution	Creative ideas for sorting device	pp. 133-137

Ideas Generated *(summarized pp. 95-97)*

Idea	Purpose	Source
Folding scoop	Pick-up device	Competing products
Conveyor with paddles	Pick-up device	Hay bale pick-up
Claw	Pick-up device	Bear's claw

Idea Selection Example for Sorters   *(others on notebook pages 156, 172, 177)*

Ideas are scored using a decision matrix as shown below. Criteria are . . .

Criterion	Importance	Scores for Proposed Sorter Concepts		
		Gravity	Shape	Density
Speed	M (2)	2	3	2
Accuracy	H (3)	4	4	4
Cost	M (2)	3	4	2
TOTAL		22	26	20

Process for Synthesis of Concept   *(see notebook pp. 198-209)*

Step	Description
1	Identify required functions for system
2	Construct means vs. function table
3	Identify interface requirements . . .
4	. . .

Process for Evaluating Solution Concept *(see notebook pp. 267-280)*

The solution concept is evaluated by (a) verifying that specs are met and (b) validating that the concept is found acceptable to users. The following methods are used for verification of specs:

Selected
Concept

Problem Definition with Specs *(see notebook pp. 165-167)*

> The solution is designed to . . .
>
> Vital specifications to be met include:
>
> | Spec Description | Targeted State | Revised |
> |---|---|---|
> | Percentage error in positioning | ±5% | 3/24/2014 |
> | Maximum cost per unit produced | $500.00 | 1/16/2014 |

Solution Concept Description

> The solution concept is . . . .
>
>   (insert photo or CAD drawing or other representation)
>
> Principal features include the following . . . with . . . unique features.
>
> Typical use of the envisioned solution is as follows:

Evidence that Specs are Met *(see notebook pp. 267-280)*

Spec Description	Targeted State	Actual	Evaluation Method
User's fatigue at 2 hrs	Acceptable	Acceptable	User validation test
Time for one cycle (sec)	<15	13.6 ± 0.7	Lab test
Power consumption (mW)	<1.5	1.32 ± 0.15	UL Std 6.13.1a

Self-Assessment of Concept

> Strengths of the solution concept include:
> - Positioning is more rapid and more accurate than existing devices
> - Disposal of spent cartridges creates no hazards to environment
>
> Concerns about the current concept include:
> - Wear on pivot needs to be tested, possibly requiring redesign
>
> Progress is slightly behind schedule. This will be addressed by . . . .

Plans for Developing Final Solution

> The following methods will be used to advance the concept to a full solution:
> - Review known concerns and make necessary revisions
> - Perform risk analysis to identify design features of importance.
> - Etc.

Scoring Sheet for Concept Generation Design Review

INSTRUCTIONS: Ask students questions listed in column 1. For each question, check off items in column 2 if they are addressed adequately by students. For each question, assign one score for the response and one for documentation supporting the response. Tally scores and circle the outcome corresponding to these scores.

Concept Generation Process

Probing Question	Addressed in Response	Acceptability Score			Documentation		
		1 Unacceptable	2 Acceptable	3 Outstanding	0 Little	1 Marginal	2 Complete
What **sources** did you use to identify ideas for your design? Describe an effective process.	□ Other designs: □ Knowledgeable people: □ Member creativity: □ Other:	Few, narrow, or irrelevant sources; vague process	Moderately relevant, varied sources; ok process	Widely varied, relevant, credible sources; strong process	Very little record; unclear on methods	Sporadic, some well-defined methods	Complete, referenced; definitive processes
Show your **results** (ideas) from idea generation efforts.	□ Large number of ideas: □ Creative ideas: □ Varied yet relevant ideas:	Few ideas; little creativity	Moderate number; some creativity	Many ideas; very creative ones; good relevance	No record of ideas	Incomplete record of ideas	Very complete record
What **process** (steps) did you use to select, evaluate and improve ideas for your design?	□ Defined process for selection: □ Consistent use of specs in evaluation: □ Parts of ideas combined with others: □ Ideas refined over time:	Ad hoc process or not by team consensus; no improvement	Moderately complete, clear process; some improvement	Very complete, clear, rigorous process; major improvement	Very little record; unclear on steps	Sporadic, some well-defined steps	Complete, , full details on steps & thinking
Explain how you **synthesized** your solution concept from component ideas.	□ Focus on whole system: □ Integration of components: □ Consider simplicity, performance: □ Consider cost, human issues:	Unable to explain; simply assembling	General steps, some system-level considerations	Focus on system integration, potential for overall success	Very little record; unclear process	Sporadic record, parts of process	Complete, record of process & thinking
Explain how you know that your selected solution **concept is "best"**.	□ Meets key specs: □ Well integrated, simple: □ Supported by data or models: □ Concept validated by users:	Vaguely justified; lacks data; lacks validation	Some evidence that specs met; parts appear integrated	Well integrated; simple; specs met per data & user confirmation	No record of concept evaluation	Incomplete record of concept evaluation	Convincing record of concept evaluation

Selected Concept

Probing Question	Addressed in Response	Acceptability Score			Documentation		
		1 Unacceptable	2 Acceptable	3 Outstanding	0 Little	1 Marginal	2 Complete
What is the problem you are solving? How has its definition been changed or confirmed?	□ Clear, compelling need: □ Potential for impact: □ Validation of need: □ Clarification of need:	Unimportant or unclear need; no confirmation or clarification	Moderate need, clearly stated; some general confirmation	Clear, urgent need; definition reaffirmed & clarified	Very little record; unclear	Single acceptable entry	Clear, refined, prominent
Show and explain the solution concept and its essential features and functions.	□ Model or representation: □ Identification of features: □ Description of functions: □ Alignment with overall need:	Unclear; poor explanation; confusing model	Good model & explanation; general fit to need	Excellent model, fit to need; and explanation	Very sketchy description & model	Adequate model and description	Excellent visuals with explanation of functions
Give evidence that the concept will meet targeted requirements.	□ Important specs: □ Analysis: □ Test data: □ User validation:	No use of specs; no solid evidence of achievement	Specs used; some data or analysis shows achievement	Key specs used; data and analysis convincing	Very sketchy records of tests or analysis	Adequate records of tests or analysis	Excellent records of process & evidence
Assess your progress, and identify strengths and areas of concern for your concept.	□ Accurate assessment of progress: □ Accurate assessment of strengths: □ Accurate assessment of concerns: □ Acceptable current state:	Inaccurate assessments, or inadequate progress	Reasonable assessments and reasonable progress	Excellent assessments and very good progress	No recorded assessment of progress or concept	Acceptable record of assessment	Clear, insightful record of assessm't
What methods do you plan to use as you develop a detailed design solution?	□ Concept is improved/built upon: □ Analysis/modeling add detail: □ Risk analysis gives focus: □ Specs/users anchor evaluation:	Unable to identify methods of value	General idea of methods for detailing solution	Clear understanding of methods and purposes	No record of plans	Vague record of plans	Specific plans defined

Concept Generation Design Review **Outcome**

Suggested Score	30-50	25-29	10-24
Outcome	Accept as is	Accept with revisions	Revise and re-review

Team Discussion: Concept Generation Design Review	To deepen your understanding and enhance your preparation for a concept generation design review, discuss the following questions within your team. Write responses for each question.

1. WHY IS A CONCEPT GENERATION DESIGN REVIEW IMPORTANT FOR YOUR PROJECT?

2. WHAT SPECIFIC THINGS WOULD YOU LIKE TO LEARN FROM A CONCEPT GENERATION DESIGN REVIEW?

3. WHAT STRENGTHS DO YOU FEEL YOUR TEAM HAS DEMONSTRATED DURING YOUR CONCEPT GENERATION?

4. WHAT SPECIFIC THINGS SHOULD YOUR TEAM HAVE DONE BETTER DURING CONCEPT GENERATION?

5. WHAT DESIGN RECORDS DO YOU HAVE TO SUPPORT YOUR CONCEPT GENERATION DESIGN REVIEW? HOW WILL YOU USE THEM IN THIS REVIEW?

6. WHAT CONCEPT GENERATION ACTIVITIES OR OUTPUTS ARE **NOT** DOCUMENTED WELL? WHAT CAN BE DONE TO FIX THIS PROBLEM?

7. GIVE EXAMPLES OF CONCEPT FEATURES THAT ARE REFINED ENOUGH TO BE USEFUL IN THE FINAL DESIGN.

8. GIVE EXAMPLES OF CONCEPT FEATURES THAT HAVE HIGH RISKS OF FAILURE OR SERIOUS CONSEQUENCES OF FAILURE, SO THEY REQUIRE MUCH MORE TESTING AND DEVELOPMENT.

5. SOLUTION REALIZATION DESIGN REVIEW

Components of Solution Realization Review

The solution realization design review is conducted after the design team has developed a design solution in full detail. Figure 6 shows the place of this design review on the overall project timeline. The solution realization process and proposed solution are expanded in this figure to show the components of each.

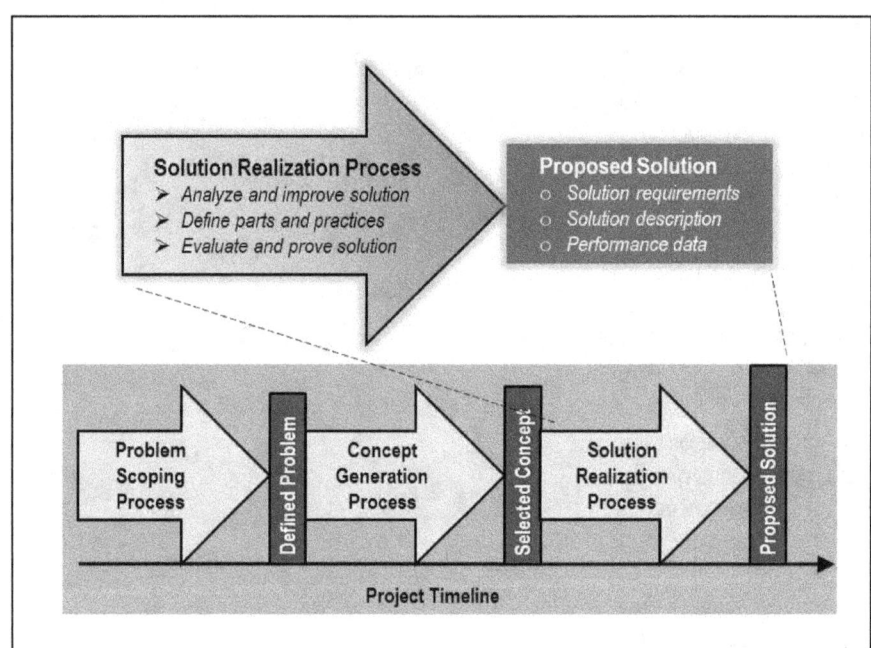

Figure 6. Components of solution realization process and products

The solution realization design review examines both the solution realization process and the proposed solution created in this design phase.

THINK: Solution Realization Review
The solution realization design review checks the quality of the solution realization process and of the proposed solution. Why is the process checked and not just the final solution? What characteristics of the proposed solution are vital before it is judged adequate to begin implementing it?
Why process:
Characteristics of proposed solution:

Adequacy of Solution Realization Process

A rigorous review of the solution realization process ensures that the proposed solution is derived from suitable problem understanding, is detailed adequately to support implementation, is assessed for risks of failure, and is analyzed and tested to ensure desired performances. Risk assessment focuses design analysis and testing on issues crucial to solution success. Final checks are made to: (a) verify that the solution meets specs and (b) validate that the solution can be applied as intended by users.

These considerations are used to define questions and scoring factors for the solution realization process review, as presented in Table 7. Questions focus on the adequacy of problem understanding, processes used to develop and refine the solution, and methods for evaluating its quality.

Table 7. Questions and scoring for solution realization process review

Question	Scoring Factors
How well validated is the problem definition?	• Effectiveness of methods to update and validate the problem definition
How adequate are actions to analyze, refine, and detail the solution?	• Effectiveness of methods of analysis • Learning and improvement evidenced • Risk assessment and reduction achieved
How adequate are actions to test, verify, and validate the solution?	• Methods for testing and analysis • Verification against specs • Validation with users

As implied by Table 7, the design team should organize the following design records to support a review of their solution realization process:

- List of efforts to update or validate problem statement and solution specifications
- List of methods used for analysis and refinement, with examples of resulting advancements (learning and design details)
- Risk assessment table showing analysis and actions
- Testing summary verifying achievement of requirements
- Summary of solution validation process

THINK: Solution Realization Process Review
The solution realization process seeks to develop the design solution to a point where it is ready for implementation. Why are methods of analysis important to this review? How can risk assessment make a solution of greater value?
Methods of analysis:
Risk assessment:

Adequacy of
Proposed Solution

A rigorous review of the proposed solution ensures that the design team and stakeholders agree that the solution meets the specifications and user expectations for its intended use. Note that the problem definition must be current, showing the best information at the time of the design review. A solution that does not satisfy solution specs will need to be revised.

Questions and scoring factors for a review of the proposed solution are given in Table 8. Questions on the defined problem ensure suitable targets for the solution, while questions on the proposed solution focus on its satisfying these requirements and applications envisioned by users.

Table 8. Questions and scoring factors for proposed solution review

Question	Scoring Factors
Briefly describe the problem and key specs for the solution.	• Clarity and importance of need/ opportunity and key specs
Describe the solution's key features and functions with evidence that these meet important specs.	• Value of solution descriptions • Evidence of achievement for (a) function, (b) financial, (c) technical, (d) safety/social requirements
Assess your solution and learning, and outline plans for solution implementation.	• Value gained in self-assessment • Merit of plans for advancing solution to its implementation

The design team should assemble the following design records to support the review of their proposed solution:

- Brief problem definition with important specs and revision dates
- Description of proposed solution with key features and functions
- Summary of evidence verifying that specs are met by the solution
- Self-assessment of proposed solution: strengths, concerns, and insights gained
- Plans for advancing solution to implementation: principal activities to be conducted to achieve various goals

THINK: Proposed Solution Review
The proposed solution represents your team's best work product to meet stated needs. Why must the team provide explicit evidence of specs achievement? What can self-assessment do for the design team?
Evidence:
Self-assessment:

Summary of Materials for Solution Realization Design Review

The following two pages contain example summaries of types of materials used to support a solution realization design review. Notebook pages point to locations of supporting materials in the engineering notebook.

Solution Realization Process

Efforts to Update/Validate Problem Definition

Date	Activity	Results	Notebook
2/7/2014	A. Tomlin & P. Diaz identified ASTM standard for ductility	Revised spec #11	p. 91
3/21/2014	Team met with P. Schwartz of Scones, Inc. to confirm specs	Specs confirmed with revision to #4	p. 129

Methods Used for Analysis and Refinement

Concern	Method	Outcomes	Notebook
Rocker arm failure	Finite element analysis	Sized rocker arm	pp.117-118
Financial returns	Fixed/operating cost analysis	Set cost limits	pp. 130-131
Linkage motion	CAD model simulation	Set arm lengths	pp. 148-149

Risk Assessment Analysis and Actions   (using FMECA)*

Potential Failure	S	L	C	RPN	Action	Notebook
Cable break (2/17/2014)	7	4	5	140	Redesign supports	p. 244
Motor burnout	3	2	2	12	Circuit breaker	p. 256
Entanglement in belts	8	3	2	40	Guard and shut-off	p. 289
Cable break (3/22/2014)	6	2	22	24	Check weekly	p. 290

*Failure Mode Effects and Control Analysis; S=severity, L=likelihood, C=control effectiveness, RPN=risk priority number (RPN=S x L x C)

Testing Summary for Component and System Requirements

Dates	Measure	Test Description	Notebook
2/3/2014	Clamping (% success)	Prototype clamp grabs blocks and delivers them to hopper (50 cycles)	p. 310
4/7/2014	Direction control (% error)	Chassis driven 10m in "forward" setting; measure error at end (10 runs)	pp. 343-344
4/9/2014	Time for repair (min)	Remove faulty switch and replace with new one (repeat 5 times)	p. 357

Summary of Solution Validation Process

Date	Process	Outcome	Notebook
2/3/2014	Team met with focus group of potential users to validate solution	0% satisfied; new concerns identified	p. 322

Proposed
Solution

Current Problem Definition with Specs *(see notebook pp. 256-257)*

> The solution is designed to . . .
>
> Vital specifications to be met include:
>
> | Spec Description | Targeted State | Revised |
> |---|---|---|
> | Percentage error in positioning | ±5% | 3/24/2014 |
> | Maximum cost per unit produced | $500.00 | 3/19/2014 |

Description of Proposed Solution

> The proposed solution is . . . .   (insert photo or CAD drawing or other representation)
>
> Principal features include the following . . . with . . . (unique features).
>
> Typical use of the envisioned solution is as follows:

Verification of Specs Achievement

Spec Definition	Target	Result*	Action	Notebook
Cycle time (sec)	<5	3.9 P	No change	p. 352
Position error (avg)	<1%	1.23% F	Refine control algorithm	pp. 367-368

* P=passed, F=failed

Validation of Proposed Solution

> Validation of the proposed solution with prospective users shows the following results:

Self-Assessment of Proposed Solution

> Strengths of the solution include:
>
> -
>
> Concerns about the proposed solution include:
>
> -
>
> Through the development of our proposed solution, the team has learned . . . .

Recommendations for Implementing the Proposed Solution

> Steps recommended for advancing and implementing the proposed solution include:
>
> -
> -

Scoring Sheet for Solution Realization Design Review

INSTRUCTIONS: Ask students questions listed in column 1. For each question, check off items in column 2 if they are addressed adequately by students. For each question, assign one score for the response and one for documentation supporting the response. Tally scores and circle the outcome corresponding to these scores.

Solution Realization Process

Probing Question	Addressed in Response	Acceptability Score			Documentation		
		1 Unacceptable	2 Acceptable	3 Outstanding	0 Little	1 Marginal	2 Complete
What **process** did you use to **update** and validate your problem definition?	□ Periodic review: □ Review as needed: □ Validated with users: □ Other:	Ad hoc process or not done	Reviewed and updated at least once	Reviewed & updated multiple times; involved users	Very little record; unclear on steps	Sporadic, some well-defined steps	Complete, , full details on steps & thinking
Explain how you used **analysis** to detail & refine your solution.	□ CAD/drawings: □ Computation: □ Physical models: □ Other:	No identifiable analysis methods; little analysis done	Identifiable methods; some engineering calculations	Multiple, defined methods, some engineering; rigorous analysis	Very little record; unclear process	Sporadic record, parts of process	Complete, record of process & thinking
Explain how you assessed and reduced **risk** as part of design solution refinement	□ Considered failure modes: □ Assessed risks: □ Attended to risky areas: □ Assessed risk reduction:	Vaguely mentioned risk; no risk analysis	Some risk analysis, but not formal or extensive	Formal risk analysis, design to reduce risk, risk reduction	No record of risk being addressed	Incomplete record of risk addressed	Extensive record of risk being addressed
List examples of **tests** done to evaluate adequacy of solution components or whole.	□ Critical issues tested: □ Sound test procedures: □ Adequate data collected: □ Other:	Few tests, inappropriate tests, or little data collected;	Some appropriate tests provide useful data	Well selected tests used effectively, yield valuable data	Little record of testing or data	Incomplete record of testing or data	Defensible record of testing & test data
How has the solution been **validated** with prospective users?	□ Appropriate users: □ Broad applications: □ Other:	No validation with potential users	Limited validation with potential users	Extensive validation with diverse users	No record of solution validation	Incomplete record of validation	Extensive record of validation

Proposed Solution

Probing Question	Addressed in Response	Acceptability Score			Documentation		
		1 Unacceptable	2 Acceptable	3 Outstanding	0 Little	1 Marginal	2 Complete
Briefly define the design problem and key specs, and explain how you know specs are right.	□ Clear, compelling need: □ Clear, relevant specs: □ Validation of specs:	Unimportant or unclear need; vague specs not confirmed	Moderately clear need;; uncertain but useful specs	Clear urgent need; specs useful and validated	Very little record of details or thinking	Moderate record of details or thinking	Complete record of details and thinking
Show and explain your proposed solution and its essential features and functions.	□ Representation: □ Key features: □ Key functions: □ Relevance to need:	Confusing or irrelevant solution; poor explanation	Good solution & explanation; general fit to need	Excellent solution and fit to need; strong explanation	Very sketchy record of solution or functionality	Adequate description of solution & functions	Excellent visuals with explanation of functions
Give evidence that the proposed solution meets a broad and important set of specs.	□ Functional specs: □ Financial specs: □ Technical specs: □ Safety/social specs:	Narrow set of specs or no solid evidence of achievement	OK range of specs; some data shows achievement	Vital specs evidenced by data and sound analysis	Very sketchy records of specs achievement	Adequate records of specs achieved	Excellent records of vital specs achieved
Assess your proposed solution (strengths & concerns), and identify design insights gained.	□ Accurate strengths: □ Accurate concerns: □ Valuable insights:	Inaccurate assessments, or shallow learning	Reasonable assessments and reasonable learning	Excellent assessments and very good insights	No recorded assessment of solution or learning	Acceptable record of solution assessm't	Excellent, record of requested assessm't
What might you recommend to any who would implement your solution?	□ Build on strengths: □ Address concerns: □ Address opportunities: □ Other:	Unable to give useful advice	Give general advice of some value	Give clear advice of real value for success	No record of advice	Scattered record of advice	Specific section for advice

Solution Realization Design Review **Outcome**

Suggested Score	30-50	25-29	10-24
Outcome	Accept as is	Accept with revisions	Revise and re-review

To deepen your understanding and enhance your preparation for a solution realization design review, discuss the following questions within your team. Write responses for each question.

1. WHY IS A SOLUTION REALIZATION DESIGN REVIEW IMPORTANT FOR YOUR PROJECT?

2. WHAT SPECIFIC THINGS WOULD YOU LIKE TO LEARN FROM A SOLUTION REALIZATION DESIGN REVIEW?

3. WHAT STRENGTHS DO YOU FEEL YOUR TEAM HAS DEMONSTRATED DURING YOUR SOLUTION REALIZATION?

4. WHAT SPECIFIC THINGS SHOULD YOUR TEAM HAVE DONE BETTER DURING SOLUTION REALIZATION?

5. WHAT DESIGN RECORDS DO YOU HAVE TO SUPPORT YOUR SOLUTION REALIZATION DESIGN REVIEW? HOW WILL YOU USE THEM IN THIS REVIEW?

6. WHAT SOLUTION REALIZATION ACTIVITIES OR OUTPUTS ARE **NOT** DOCUMENTED WELL? WHAT CAN BE DONE TO FIX THIS PROBLEM?

7. GIVE EXAMPLES OF SOLUTION FEATURES THAT ARE REFINED ENOUGH TO BE USEFUL IN THE IMPLEMENTATION OF YOUR PROPOSED DESIGN.

8. GIVE EXAMPLES OF SOLUTION FEATURES THAT HAVE HIGH RISKS OF FAILURE OR SERIOUS CONSEQUENCES OF FAILURE, SO THEY REQUIRE MUCH MORE TESTING AND DEVELOPMENT.

APPENDIX:

TEMPLATES FOR DESIGN DOCUMENTATION

Templates for Documentation

Documentation of design should be an ongoing process that occurs as design activities are planned and executed. Documentation should capture goals of the design work, activities and accomplishments, design products, questions and thoughts that arise, and insights gained from the work. No single format fits all situations and documentation needs.

Table 9 identifies templates found on the next several pages that may be used to document design activities and outcomes.

Table 9. Templates for documenting design activities and outcomes

Title	Purpose
Work Session Record	Document meetings or distinct periods of individual or team design activity
Information Gathering Record	Document process, sources, and findings from information gathering activity
Solution Requirements Record	Document problem statement and solution specifications at a distinct point in time
Idea Generation Record	Document process, sources, and ideas generated for a specific design challenge
Idea Screening Record	Document qualitative method used and ideas screened for a specific purpose
Concept Selection Record	Document quantitative scoring process used to select concept best meeting criteria for a design component
Selected Concept Record	Document features and functions of a component or full solution concept at a distinct point in time
System-Level Concept Selection Record	Document selection of means for achieving desired functions in a system-level design
Sub-System Interface Requirements Record	Document relationships among solution sub-systems that establish requirements for full system design
Performance Testing Record	Document testing procedure, data collected, and results that evaluate performance of solution
Current Design Solution Record	Document a summary of features and functions of the design solution at a distinct point in time
Solution Verification Record	Document achievement of specifications by the design solution at a distinct point in time
Solution Validation Record	Document procedure and outcomes from user application of solution to meet specific needs

Work Session Record

Date: Location: Recorder:

Participants:

Goal(s) for Session:

Tasks to be Done

Task	Responsible Person(s)
1:	
2:	
3:	
4:	

Summary of Outcomes Achieved

Task	Outcomes
1	
2	
3	
4	

Reflections

Strengths

Concerns

Insights

Information Gathering Record

Date: Participants: Recorder:

Sources Used
(check all that apply):

☐ Text books   ☐ Websites   ☐ Patents   ☐ News articles   ☐ Technical reports   ☐ Market studies

☐ Surveys   ☐ Focus groups   ☐ Product literature   ☐ Product testing   ☐ Personal experience

☐ Experts   ☐ Project sponsor   ☐ Other (specify):

Goal(s) for Information:

Sources and Findings

Specific Source	*Information Found*
1:	
2:	
3:	
4:	
5:	

Reflections

Strengths	
Concerns	
Insights	

Solution Requirements Record

Date: Contributors: Recorder:

Project Title:

Problem Statement

Need or Opportunity	

Envisioned Solution	

Anticipated Benefits	

Solution Specifications

No.	Description of Specification	Targeted State	Importance (H/M/L)	Date Updated	Source
1					
2					
3					
4					
5					
6					
7					
8					
9					
10					
11					
12					
13					
14					
15					
16					
17					
18					
19					
20					

Idea Generation Record

Date: Contributors: Recorder:

Project Title:

Purpose of Ideas:

Process Used to
Generate Ideas:

Generated Ideas and Sources

Description of Idea Source	Sketch of Idea

Idea Screening Record

Date: Contributors: Recorder:

Project Title:

Purpose of Ideas:

Criteria Used to
Screen Ideas:

Ideas Screened by Relevant Criteria

Score each idea for fit to each criterion (e.g., speed, cost) using scores of: 2=good, 1=ok, 0=bad

Description of Idea	Criterion 1:	Criterion 2:	Criterion 3:	Criterion 4:	Criterion 5:	Total Score

Outcomes

Decisions

Insights

Concept Selection Record

Date: Contributors: Recorder:

Project Title:

Purpose of
Concept:

Selection Criteria Definitions

Criterion Name	Criterion Definition

Concept Scoring by Selection Criteria

Assign each concept a weight (1 to 3) for importance, then score it 1=poor to 5=excellent for fit to each criterion.

Description of Concept	Weight (1 to 3)	Scores (1 to 5) for Concept Fit to Criteria					Weighted Total Score
		Criterion 1:	Criterion 2:	Criterion 3:	Criterion 4:	Criterion 5:	

Outcomes

Decisions

Insights

Selected Concept Record

Date: Contributors: Recorder:

Project Title:

Purpose of
Concept:

Definition of Selected Concept

Concept Visualization	*(insert here a photo, drawing, diagram, or other visual representation of major elements of the concept)*
Description of Principal Features	*(list here the important feature of the concept that make it attractive as a starting point for a detailed solution)*
Definition of Essential Functions	*(list here the important functions the concept will perform to satisfy requirements for a successful solution)*

Self-Assessment of Concept

Strengths

Concerns

Insights

System-Level Concept Selection Record

Date: Contributors: Recorder:

Project Title:

System:

Essential Functions of System

Function Name	Function Definition

Morphological Chart (Means of Fulfilling Functions)

Function	In each cell of a row, define a means (concept) for achieving the designated function. Select (circle) the best combination of means to achieve all functions in a compatible way.					

Self-Assessment of System Concept

Strengths	
Concerns	
Insights	

Sub-System Interface Requirements Record

Date: Contributors: Recorder:

Project Title:

System:

Sub-System Definitions

Sub-System Name	Sub-System Description/Functions
Example: Power system	Power from engine energizes drive train, lifting mechanisms, sensors, and controls

Sub-System Interface Requirements

Sub-System Name	Interface Requirements with Other Sub-Systems
Example: Power system	• Power transmitted to drive train by rotating motion from engine; disengaged by clutch; wheel rotation speed ranges from 1/10 to 4/10 of motor speed; safety lock-out required • Power transmitted to sensors by engine-driven 9v electric generator . . .
	•
	•
	•
	•

Reflection on Sub-System Interfaces

Performance Testing Record

Date: Contributors: Recorder:

Project Title:

Component being Tested:

Question Asked:

Measurement(s)   •
to be Made: •

Statistical Test(s)
to be Made:

Experimental   1.
Steps for Data   2.
Collection   3.
                      4.
                      5.

Data Collection

Run No.	Measure 1:	Measure 2:	Measure 3:	Measure 4:
1				
2				
3				
4				
5				
6				
7				
8				
9				
10				

Statistical Analysis

Conclusions (confirmations, concerns, insights)

Current Design Solution Record

Date: Contributors: Recorder:

Project Title:

Problem Summary
- Overall need
- Solution desired
- Benefits desired

Definition of Current Solution

Solution Visualization	*(insert here a photo, drawing, diagram, or other visual representation of the current design solution)*
Description of Principal Features	*(list here important features of the solution that make it valuable; note any that are novel)*
Definition of Essential Functions	*(list here important functions the solution performs to satisfy stated requirements)*

Assessment of Current Solution

Self-Identified Strengths or Concerns	
User-Identified Strengths or Concerns	

Solution Verification Record

Date: Contributors: Recorder:

Project Title:

Project Status
(check all that are
completed):

☐ Solution concept defined   ☐ Major components defined   ☐ Components manufactured or purchased

☐ Risks assessed   ☐ Components revised   ☐ Component function evaluated   ☐ Overall system evaluated

Verification of Specifications Achievement

Spec Definition	Targeted State	Date Defined	Date Tested	Actual State	Outcome (pass/fail and action)	Notebook Pages

Assessment of Solution Verification

Strengths or Concerns	
Action Needed	

Solution Validation Record

Date: Contributors: Recorder:

Project Title:

Project Status
(check all that are
completed):
  □ Solution concept defined   □ Major components defined □ Components manufactured or purchased

  □ Risks assessed □ Components revised □ Component function evaluated □ Overall system evaluated

Validation that Solution Fulfills User Expectations

User Name (and Group)	Description of Application	Outcomes of Testing (strengths, weaknesses)	Notebook Pages

Assessment of Solution Validation

Strengths	
Concerns	
Follow-up Actions	

www.ingramcontent.com/pod-product-compliance
Lightning Source LLC
Chambersburg PA
CBHW081638220526
45468CB00009B/2481